Solomon Jeremiah

Environmental Impacts of Round Potato Production at Kibohelo Hamlet Lushoto District, Tanzania

GRIN Verlag

Bibliografische Information der Deutschen Nationalbibliothek:

Die Deutsche Bibliothek verzeichnet diese Publikation in der Deutschen National-
bibliografie; detaillierte bibliografische Daten sind im Internet über http://dnb.d-
nb.de/ abrufbar.

Impressum:

Copyright © 2015 GRIN Verlag GmbH
Druck und Bindung: Books on Demand GmbH, Norderstedt Germany
ISBN: 978-3-656-94288-7

Dieses Buch bei GRIN:

http://www.grin.com/de/e-book/296098/environmental-impacts-of-round-potato-
production-at-kibohelo-hamlet-lushoto

GRIN - Your knowledge has value

Der GRIN Verlag publiziert seit 1998 wissenschaftliche Arbeiten von Studenten, Hochschullehrern und anderen Akademikern als eBook und gedrucktes Buch. Die Verlagswebsite www.grin.com ist die ideale Plattform zur Veröffentlichung von Hausarbeiten, Abschlussarbeiten, wissenschaftlichen Aufsätzen, Dissertationen und Fachbüchern.

Besuchen Sie uns im Internet:

http://www.grin.com/

http://www.facebook.com/grincom

http://www.twitter.com/grin_com

ENVIRONMENTAL IMPACTS OF ROUND POTATO PRODUCTION AT KIBOHELO HAMLET LUSHOTO DISTRICT, TANZANIA

By

Sembosi Solomon Jeremiah

Department of Ecotourism and Nature Conservation, Sebastian Kolowa Memorial University,

P. O. Box 370, Lushoto, Tanzania

ABSTRACT

This study seeks to document the Environmental Impacts of round potato production at Kibohelo hamlet, Lushoto district, Tanga Tanzania. The study was conducted in one hamlet, namely Kibohelo hamlet, as the representative of the whole Lushoto district and Tanga region in general as one of the area where round potato are being produced. The study evaluates the awareness of local community on the environmental impacts of potato production where by findings from this study indicate that majority of respondents, (53.57%) were aware of the changes in environment around cultivated field. However, (21.43%) respondents claimed not to know anything about environmental impacts of potatoes production or any changes related to it. The study also identifies farming practices whereby crop rotation (28.57%), mulching (21.43%), miraba (16.07%) and small scale irrigation (19.64%) were found to be used by the farmers. As well as determines the problems associated with the farming practices to the environment whereby soil by (55.36%) as commented by farmers in which its perceived to have changed than the other environmental aspects this is due to application of fertilizers, pesticides, herbicides and chemicals which are used to remove fog over the crops which directly end up to the soil. The other aspect seeming to have change is water, where by 1.79% of the farmers reported it. Purposive and Simple random sampling procedures were employed to collect the sample with the sample size of 56 respondents from the study area. Data were collected through the main technique including interviews based upon questionnaire forms and personal observation through observation guide were conducted and the data collected from questionnaire were analyzed by computer software known as SPSS. The cross tabulations, charts, frequencies and percentages were obtained whereby results are presented in form of statement, tables, charts or graphs. The findings of this study are crucial for the creation and increase community's awareness (especially farmers) as well as the results are useful to the Ministry Of Agriculture, Food and Cooperation to prepare and or in the innovations of agricultural policies and plans about adaptation and mitigation measures to deal with the problems associated with the farming practices to the environment but also the study adds more knowledge on the existing literature and will act as supportive insights for further researches.

Key words: round potato, farmers, environment, impacts, Tanzania, developing countries

INTRODUCTION

Potato (solanumtuberosum) is the world's most important root and tuber crop worldwide. It is grown in more than 125 countries worldwide and consumed almost daily by more than a billion people. Hundreds of millions of people in developing countries depend on potatoes for their survival. Potato cultivation is expanding strongly in the developing world, where the potato's effort of cultivation and nutritive content has made it a valuable food security and cash crop for millions of farmers.

Potatoes rank fourth in the world as a food crop after maize, rice, and wheat (FAOSTAT, 2009) and have been recognized as one of the main crops to alleviate hunger in the world. Toulmin (2011) pointed "Feeding the world sustainably and more fairly requires us to overcome several substantial hurdles since the world's population is expected to grow by nearly a third by 2050, and more food will be needed and we shall expect more problems than today.

The top five potato producing countries in the world includes, China (74,799,084 m/t in 2010 and 73,281,890 m/t in 2009), India (36,577,300 m/t in 2010 and 34,390,900 m/t in 2009), Russian Federation (21,140,500 m/t in 2010 and 31,134,000m/t in 2009), Ukraine (18,705,000 m/t in 2010 and 19,666,100m/t in 2009) united states (18,016,200m/t in 2010 and 19,564,300 m/t in 2009) (FAOSTAT data, 2012). Were some environmental problems have been reported such as water contaminations (pollution) and soil pollution.

Developing countries are now the world's biggest producers and importers of potatoes and potato products and Potato production in Africa is dominated by four countries which are Egypt, South Africa, Algeria and Morocco (Alcamo, J. 2001).

In Tanzania, potatoes are sometimes called "Irish potatoes" or "European potatoes" (ViaziUlaya in Swahili), indicating their foreign origin, or "round potatoes" (Viazimviringo in Swahili) distinguishing them from sweet potatoes (Andersson, 1996).

This means that round potato can address both food security as well as profit. Indeed this round potato is more profitable than traditional staples, as it has higher yield per unit of land, matures earlier, and provides a larger income (Blanken*et al.,* 1994; CIP, 2008). The maturity period of round potato is about three months as compared to maize (the major staple in Tanzania), which takes about eight to ten months to be ready for harvest (UARC, 1990). Also, one acre of round potato produces up to 120 (100 kg) bags versus about 20 bags of maize on the same piece of land in addition the price per unit are comparable although round potato sales are often higher than maize (BOT, 2010; UARC, 1990). For the round potato, these data show that it is a good income earner, and because of its potentiality, the crop is considered to be a hidden treasure for smallholder farmers (Blanken*et al.,* 1994; CIP, 2008).

Apart from its contribution to food security and economic earnings potato has got some environmental impacts or problems which associate with it particularly during its production. The problem includes emission of greenhouse gases and leaching of used fertilizers. On the other hand the use of soluble chemical fertilizers for crop production particularly to supply nitrogen, phosphorous and potassium has increased potato yields and quality for several decades. Over the past 10 years there has been an increased concern over the environmental impacts of agricultural fertilizers, particularly as nonpoint source of water pollution. Currently new challenges are arising such as concerns over phosphorus leaching and heavy metals contamination in agricultural fertilizers. These have the potential to restrict nutrient use in agricultural systems, requiring both potato producers and scientists to seek additional alternatives to improve nutrient use efficiency. Whereby during cultivation, it is the use of fertilizer's particularly mineral fertilizers and ploughing that releases greenhouse gases (Wichelns, D., & Oster, J. D. 2008).

On the other side of using pesticides in growing potatoes it depends on the quantities of pesticides that are used in the cultivation, how they are handled and the type of pesticides, which in turn, depends on the cultivation technique and climate factors Example In areas with cold climate, the problem of pests and diseases is often less significant, so that in general smaller quantities of pesticides are used than in warmer areas (Williamson et *al.,* 2008).As the global demand for potato increases, finding ways to grow more potato while preventing environmental degradation and insuring environmental safety will be essential to help ensuring food security.

Statement of the problem

Agriculture is by far Tanzania's most important economic sector, in terms of both employment provision and contribution to Gross domestic product (GDP) (Andersson, 1996). Unfortunately, the large degree of dependency on this sector renders the environment to more problems associated with the farming practices within the environment.

The general trend towards more intensive and industrialized agriculture as one of the farming practice in many places has a profound impact on the environment, including emissions to air Example during expansion of agricultural land (example converting forest to crop land) can increase greenhouse gas emissions example carbon dioxide (Co_2), whereby currently it's estimated that 75% of nitrous oxide (a gas with a global warming potential that is 310 times greater than carbon dioxide) is produced from agriculture, whereby approximately 60% is produced directly from agricultural soils. The husbandry factors that are associated with high nitrous oxide emission include high rates of nitrogen fertilizer, either as manure of manufactured product. Typically, emissions of nitrous oxide from potato crops are higher than from small-grain cereals or oilseed rape crops. But also quality and quantity of surface water and groundwater due to leachate from agricultural lands, soil erosion, pollution due to high use of pesticides leading to loss of biodiversity and habitats due to lack of conducive environment for growth and thrive (EEA 2006a, EEA 2006b).

The production of potatoes in Magamba at Kibohelo is ongoing cultivation process however much of its impact on the environment is not well documented locally and therefore the magnitude of its impact is as well not well known by both farmers and extension staff members. Regarding that in cultivation of potatoes farmers make use of synthetic fertilizers and pesticides, it is on the basis of this study that was carried out to explore the problems associated with the farming practices to the environment at Kibohelo hamlet in Lushoto District with a focus to potato production in both rain fed and irrigated production systems, the results which will be extended to other areas for better production and environmental management.

Objectives

1.2.1 Overall objective

The overall objective of this study is to establish information on environmental impacts of potato production in Kibohelo.

1.2.2 Specific Objectives

The specific objectives include the following;

i) To identify the awareness of local community (Farmers) on the environmental impacts of potato production.

ii) To identify the sources of information on environmental impacts of potatoes production.

iii) To identify the farmers perception on environmental Impacts of potatoes production.

iv) To identify farming practices, trends, use of improved seeds, pesticides and the existence of value chain in potatoes production at the kibohelo hamlet

1.3 Research questions

    i)   What are the farming practices done within the area?

    ii) What are the environmental problems associated with the farming practices within the area?

    iii) What correlations exist between the farming practices and problems to environment?

    iv) Are the local communities aware of the environmental impacts of potato production?

1.4 Importance/significance of the study

The results of this study are expected to assist the following categories of people/institutions in making better adoption plans and actions in view of the problems associated with the farming practices to the environment.

i.   To the local communities; the study will create and increase community's awareness (especially farmers) on the environmental impacts associated with farming practices and it will enable them to plan adaptations measures that will cope with healthier environment while enhancing the increase in production yields.

ii.   To researchers; the study will add knowledge to researchers that will create and increase awareness on the problems associated with the farming practices to environment so that they can come up with solutions to address them.

iii. To The Government through the respective Ministry of Agriculture, Food and cooperation as well as planners and policy makers. The results obtained from this study will be useful to the Ministry Of Agriculture, Food and Cooperation to prepare and or in the innovations of agricultural policies and plans about adaptation and mitigation measures to deal with the problems associated with the farming practices to environment in Lushoto District as a whole.

RESEARCH METHODOLOGY

Description of the study area

Location

The study was conducted in Lushoto district particularly in the valley adjacent to the hamlet Kibohelo in Magamba village, which is located on the road from Lushoto to Lukozi. This valley has been selected because of its wide variety in landscape elements and land use types. Some of the land use types encountered within the area are natural forest, plantation forest, rain fed and irrigated agriculture especially on vegetables and round potato in the valley bottom.

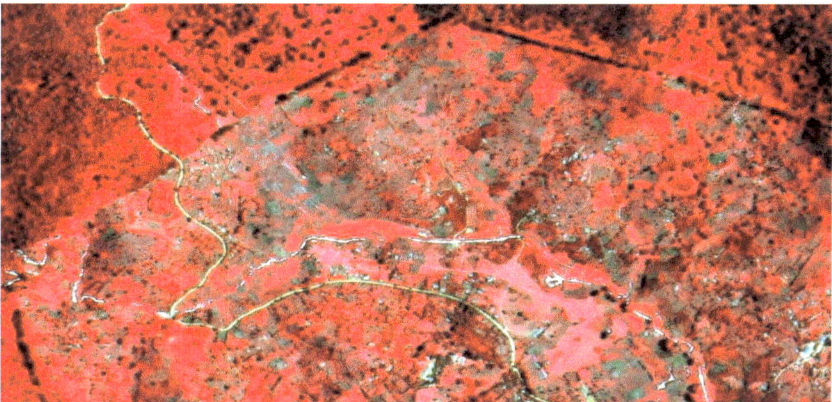

Figure 2 Spot image of kibohelo village, Lushoto District Tanzania.

Lushoto district is situated in north eastern Tanzania within latitudes 4^0 22' to 5^0 08 and longitudes 38^0 5 to 38^0 38'', Lushoto district covers 3500 km^2 out of which 2000 km^2 are arable (Mowo et al., 2002). The topography of Lushoto is mountainous and its altitude ranges from 600 meters to 2300 meters above sea level. The geological materials from which the soils of Lushoto are derived include the metamorphic rocks like schists and gneisses. These cover the bulk of the district (Ngailo et al., 1998). Meliyoet al., (2002) have indicated the major soil types in Lushoto as being Humic and Chromic Acrisols, Luvisols and Lixisols for most of the mountainous highlands' while the valley bottoms have Fluvisols and some pockets of Gleysols. Rainfall in Lushoto is bimodal and the precipitation varies from 900 to 1300 millimeters per annum. Figure 2 shows rainfall pattern in Lushoto for the last 70 years.

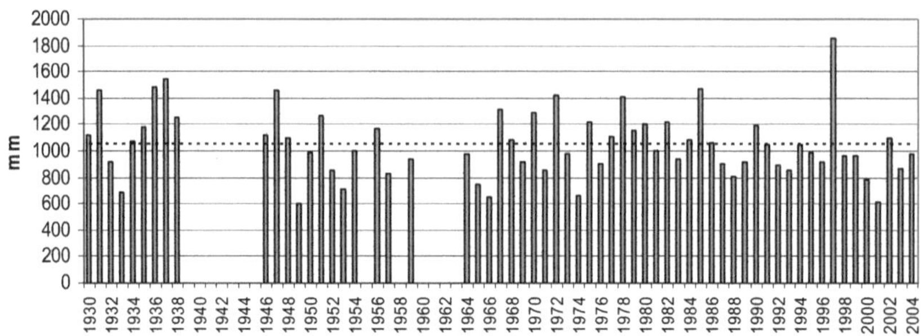

Figure 3 Rainfall pattern of Lushoto (70 yrs)

According to Pfeifer (1990) rainfall in Lushoto is greatly influenced by the agro-ecological zone of an area. Pfeifer reports four agro-ecological zones (AEZ) for the district: The Humid Warm Zone; The Dry Warm Zone; The Dry Cold Zone and the Dry Hot Zone.

The mean annual temperature varies with altitude that is, at 500 m.a.s.l the range is 25-27°C while on the plateau, at 1,500-1,800 m.a.s.l. the range is 17-18°C (Mettrick, 1993).

3.1.2 Natural Vegetation

Within the plateau the Montane rain forest covers the area and contains dominant species like Juniperousprocera, Podocarpususambarensis, and Ocoteausambarensis. Towards lowlands there is dry forest with few trees and shrubs but towards the east there is small portion of closed lowland rain forest. The lowland is dominated by species like Khayaanthoteca, Pterocarpusholtzii, Makhamialutea and Terminaliasambesiaca (Kato, F. 2007).

3.1.3 Economic activities

The majority of Lushoto district dwellers are engaged in artisan agricultural activities with maize and beans being a staple food and horticultural crops such as potato, cabbage, tomatoes and carrots as cash crops (Miguel, 2004). The Lushoto area is very suitable for growing fruits which are harvested and transported to outside markets such as Dar es Salaam, Moshi and Arusha (Mettrick, 1993).

8

3.1.4 Population and ethnicity

Lushoto accounts for nearly 24.1% of Region's total population but the exact population from 2012 population census indicated that total population of Lushoto district is 492,441 out of which 230,336 (46.8%) are males and 262,205 (53.2%) are females (Masija, 1983). The sambaa are the most predominant group in Lushoto district making 78% of the total population, whereas pare, 14% and the mbugu 8% are minority (Masija, 1983).

3.2 Research Design

This study used cross sectional research design. This design was chosen because it studies a cross-section of the population at a single point in time, and data collection is done once. Therefore due to a limited time for conducting the study, this designwas supportive in saving time as very big sample was used in data collection. This design was also helpful to this research in obtaining accurate and relevant results since it employ the use of different methods of data collection such as observation, questionnaires and check list. In which as a researcherwas able to talk with the respondents and prove some of the answers given by respondents by observation.

3.3 Sample size and sampling procedures

Lushoto has a total of 8 divisions 32 wards and 176 villages. Purposive sampling has been used to choose Lushoto district because it is one of the areas where potato production is one of theeconomic activity therefore it enabled the research inmeeting the objectives of the study. Magamba village particularly kibohelo hamlet was also selected by using purposive sampling because they practice Irrigated potato production and rain-fed potato production respectively. Therefore it enabled the researchto determine the environmental impact of potato production both under irrigated and rain-fed potato production systems.

3.3.1 Sample size

A simple random sampling was used to select households to be surveyed and be interviewed in the area. A total of 56 respondentswere interviewed. Also random sampling was useful because it was providing equal chance for the households to be selected. Whereby, Kibohelo hamlet has the total population of 505 people (population censer 2012). Since due to time and financial factor I was unable to work with the whole population and therefore I took the sample to represent the whole population. The sample size selected was obtained by the following formula and how it was calculated.

$$n = \frac{N}{1 + N(e)^2}$$

(Kothari, 1985)

Where:

n = sample size, N = total population of the village and e = level of confidence

Therefore, n=?

N= 505 and e = 5% = 0.05

$$n = \frac{505}{1 + 505(0.05)^2}$$

$$n = \frac{505}{2.2625}$$

$$n = 223.2 = 223$$

Therefore the sample size (n) used was 223 respondents because the decimal number is 2 so it was be rounded up. But from this sample size only 25% of respondents were interviewed due to scarcity of time and financial resources. The 25% of the sample size was calculated as;

$$\frac{25}{100} \times 223 = 55.75 = 56$$

Therefore my sample size was 56 respondents.

3.4 Data collection procedure

Quantitative and qualitative data for this study were obtained through primary and secondary sources. In addition, during the survey, the physical observations was made and where applicable the photographs were taken.

3.4.1 Primary data collection

Primary data are the fresh data collected for the first time from the field. It was collected by the use of questionnaires, checklist and observation method.

3.4.1.1 Household questionnaire

The method was helpful to extract information from the household. Both closed and open questionnaire were used during interviewing process. Household questionnaire were employed to provide key information on the problem under study because household are the one who are practicing the potato production.

3.4.1.2 Checklist

The method wasuseful to get information from key informants like village government leader and agriculture extension officers. Key informants refer to any individual who is accessible, willing to talk and have a great depth of knowledge about issue in question.

3.4.1.3 Physical observation

This technique was used throughout the study for questions which the researcher can answer by himself. The method of researcher observation will be primarily used to tie together the more discrete elements of data gathered by other methods.

3.4.2 Secondary data

Secondary data were obtained from Lushoto Farmers Training Centre and internet. Other sources included SEKOMU library, books, publications and journals. Some information were obtained from the District offices and farmers who had information relevant to the study.

3.5 Data processing and analysis

The data passed through 3 processing levels that are editing, coding and actual analysis, after which a presentation of the data followed.

3.5.1 Data Editing

The data collected fresh from the field were assessed to detect errors, omissions, contradictions and unreasonable information to be corrected. This was done to ensure that data are accurate, consistent, uniformly entered and well arranged to facilitate coding and tabulation.

3.5.2 Data coding

This is an exercise that was carried out after the data collected have been edited, where numerals and other symbols were assigned to questionnaires for easy entry of data in the computer software for the actual analysis.

3.5.3 Data analysis

After the data had been edited and coded, then they were entered into the computer for analysis. The actual analysis was done using Statistical Package for Social Science (SPSS) programmeand Microsoft office Excel. The computer software of SPSS was used in the actual analysis of the collected data and it was useful in analyzing data by their percentages, frequencies and in forms of tables, charts and graphs.

3.5.4 Data presentation

When the data were already analyzed then data presentation followed in which the analyzed data were presented in form of charts, graphs, tables and figures.

RESULTS AND DISCUSSION

Socio-economic characteristics of respondent

Gender of respondents

Results in **Table 1** shows that both men and women were involved in potatoes production in the village (Kibohelo). Men accounted for about 60.7% and women about 39.3%, indicating most of the households were headed by men and thus why they are highly involved in potatoes production than womenbut also due to few activities to engage with in the village.

The findings by Duze and Mohammed 2006 also document that, men are involved more than women in agricultural activities and this is mostly influenced by few activities that men can have in rural areas to earn for sustaining their families, therefore for the case of the impacts caused by potato production, men involves more than women (Duze and Mohammed, 2006).

Table 1: Gender of the respondents

Gender of the respondents	Kibohelo	
	n	%
Male	34	60.7
Female	22	39.3
Total	**56**	**100**

Source: Field data 2014

4.4.2 Age of respondents

Regnard (2006) urges that the total accumulation of wealth is highly dependent on age of an individual, whereby a direct relationship is experienced. Likewise, age determines individual maturity and ability to make rational decisions. Moreover, Mlambiti, (1994) shows that age structure can be used to facilitate an understanding about labor potential of a specific population and such theories could be applied even in observations of age respondents involving in agriculture activities in rural areas. Results in **Table 2** give a summary of ages of Kibohelo farmers. Majority of the farmers (42.9%) were at the age 31 to 41. Indicating that most of the respondents were of the middle age, the age at which they are still energetic and hence can actively involve in production activities, such could also be confirmed by the findings by Lupilya, 2007.

Table 2: Age of respondents

| | Kibohelo | |
Age (years)	N	%
Less than 15	0	0
20 – 30	16	28.6
31 – 41	24	42.9
42 – 52	10	17.9
53 – 63	2	3.6
64 – 74	4	7.1
Total	**56**	**100**

Source: Field data 2014

4.4.3 Marital status of respondents

The study results identified that majority of farmers in Kibohelo village (82.1%) were married. While divorced respondents could not be documented throughout the study. About 17.9% of respondents were still single (**Table 3**). This indicates that most people who engage in potatoes production in the village produced potato to meet the need of their families.

Findings by Kisinza*et al.,* 2008 and Matilda 2010 also indicates that most of the people in rural areas engage in agriculture as their main activity to earn their basic needs and as a source of income.

Table 3: Marital status

| | Kibohelo | |
Marital status	N	%
Married	46	82.1
Single	10	17.9
Divorced/separated	0	0
Total	**56**	**100**

Source: Field data 2014

4.4.4 Education level

Existing literatures show that education contributed 50% of variation in the total agricultural output in Tanzania (Amani *et al.,* 1989). **Table 4** gives a summary of education levels of the respondents in the village. Results show that majority of farmers (50.0%) acquired primary education while just 30.4% of farmers have attained secondary education level, about 19.6% did not attend school and no one who have attend the college. This implies that farmers have low basic knowledge concerning agronomic skill that can be used to improve potatoes production.

The low level of education makes it difficult for this farmers to know the consequences of the agricultural methods they use, Example over application of synthetic fertilizers and pesticides with the intent of increasing yields without knowing the impacts they brought to the environment (Junge*et al.*, 2009).

Table 4: Education Level

Status	Kibohelo	
	n	%
Did not attend school	11	19.6
Primary school	28	50.0
Secondary school	17	30.4
College	0	0
Total	**56**	**100**

Source: Field data 2014

4.5Awareness on Environmental impacts of potatoes production at Kibohelo village.

Findings from this study indicate that majority of respondents, (53.57%) were aware of the changes in environment around cultivated field. This observation is highly supported by reasons that majorities of the farmers have presently observed on how the main aspects of the environment such as soil and water have negatively been affected as compared to previous years as Figure 4 indicates. However, (21.43%) respondents claimed not to know anything about environmental impacts of potatoes production or any changes relate to it (Figure 3).

From other studies elsewhere reported that, lack of accessibility to extension and credit services in many parts of Sub-Saharan Africa as well as other developing countries is the limiting factor for increased awareness among farmers and this has result to more impacts from agricultural activities. (Eze *et al.*, 2006; Junge *et al*, 2009; Okoedo-Okojie and Onemolease, 2009).

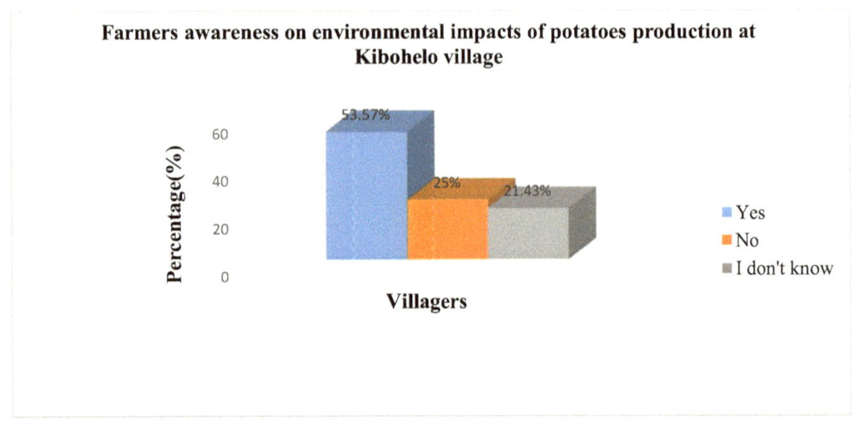

Source: Field data 2014

Figure 4: Farmers' awareness on environmental impacts of potatoes production in the village.

4.6 Sources of information on environmental impacts of potatoes production

Eze *et al.,* 2006 and Prakash, A. 2010indicates that lack of reliable information from different sources to the farmers brought unfamiliarity of their influences within the environment and thus good reliable as well as applicable information to the farmers which must go together with educating them, will help to rise their awareness pertaining environmental impacts of their agricultural practices.

Results of this study indicate that, while majority of the respondents 46.4 % in the village were told by the researchers about the environmental impacts of potatoes production, and 19.6% claimed that they observed the changes by themselves. While 10.7%of the respondents obtained information on the impact of potato production from village meetings. The least used sources of information were televisions, NGO's, being told by neighbors, friends or family which accounts for 7.1% and only 1.8 % of the respondents obtained information from input suppliers of fertilizers and pesticides, while 14.3 % of respondents were unaware of impacts that potato production have on environment. There is a room to improve engagement of unpopular sources such as NGOs and input suppliers concerning environmental changes due to potatoes production information dissemination. Furthermore based on the level of awareness some of the respondents 10.7% were very well informed on environmental impacts of potatoes production, while majority of the respondents 35.7% claimed to be fairly bad informed and 33.9% claimed to be informed fairly well. Summary of these is shown in **Table 5**.

Table 5: Source of information concerning environmental impacts of potatoes production

Source of information	Kibohelo	
	N	%
Own Observation	11	19.6
Village meetings	6	10.7
Told by neighbors, friends or family	4	7.1
Input suppliers	1	1.8
Researchers	26	46.4
None	8	14.3
Total	**56**	**100**
How informed the farmers are :-		
Very well	6	10.7
Fairly well	19	33.9
Fairly bad	20	35.7
Very bad	11	19.6
Total	**56**	**100**

Source: Field data 2014

Moreover, the results shows a remarkable observation that 36 (64.3%) respondents reported to have heard or experienced the changes in environment pertaining potatoes production only recently compared to 12 (21.4%) who have heard or experienced long ago that is since five years back. However 8 (14.3%) respondents never heard about impacts of potatoes production to the environment. Reason for this observation could be that the majority of the interviewed respondents were not elderly who justifies their experiences (indegeneous knowledge) in tracking the changes in the environment in their respective areas as **Table 6** indicates.

But also low or lack of education, lack of extension services, lack or poor farm record keeping by farmers also inhibit awareness of the changes within the environment. Minoe *et al.,* (2003) and Prakash, A. 2010noted that lack of keeping farm records by farmers is more pronounced in subsistence economy due to high levels of illiteracy in most low resource African farming countries.

Table 6: When the farmers heard about Environmental Impacts of potatoes production

When farmers heard about Environmental Impacts of potatoes production	Kibohelo	
	N	%
Recently	36	64.3
Long ago	12	21.4
None of the above	8	14.3
Total	**56**	**100**

Source: Field data 2014

4.7 Farmers Perception on Environmental Impacts of potatoes production in the village

4.7.1 General perception on Environmental Impacts of potatoes production in the village

In general, as shown in **Table 7** Environmental Impacts due to potatoes production was perceived to be a bad thing by farmers as reported by 56 (100%) respondents. Reasons behind these observations could be due to the fact that they have seen the impacts to have caused changes to the soil quality which in turn led to crop failure, poor yield, loss of income and hunger at some instances. Respondents also recalled and said twenty years back in their village when soils were highly fertile they used to grow potatoes with no diseases and no need for much manure application and sometimes no need for its application at all.

Gigliotti, C., Allievi, L. (2001), document that perceived severity of the negative impacts caused by the changes in environment due to potato cultivation underlies the knowledge on awareness and understanding of the impacts caused by potato cultivation and its related risks as well as the ability of a farmer to cope with the impacts he/she is exposed to.

Table 7: Farmers' perception to Environmental Impacts of potatoes production in the village

Farmer's perception to Environmental Impacts of Potatoes production	Kibohelo	
	n	**%**
Bad	56	100
Good	-	-
I don't know	-	-
Total	**56**	**100**

Source: Field data 2014

Generally, within the study area the aspects of the environment were perceived to have change, where by majority of the farmers (55.36%) commented that, soil is perceived to have changed than the other environmental aspects this is due to application of fertilizers, pesticides, herbicides and chemicals which are used to remove fog over the crops which directly end up to the soil, this similar observation was reported by UARC 1990 in Mbeya district Tanzania. The other aspect seeming to have change is water, where by 1.79% of the farmers reported it, and it's due to cultivation near the Kibohelo river where by much of the leachate from the fields ends up in water as well as some of the containers which are used, are washed in the water stream although it is prohibited for the safety of the villagers who are using water from the river Kibohelo and this is due to some cases reported by some villagers who were affected after using the water from the river and get stomach problems and allergies due to dissolved chemicals in the water. However 39.29% of the villagers reported that both soil and water have changed and 3.75% said they don't know particularly due to unawareness of the impacts brought by potatoes cultivation.

Defra (2009),Minoe*et al.,* (2003) and Gordon A (2010), indicates diffuse pollution from agricultural land occurs when nutrients in the soil, pesticides, sediment (soil particles) and pathogens (e.g. *Cryptosporidiaspp)* are washed from the land and into watercourses or sub surfaces. The potato crop typically receives large inputs of the plant fertilizers, nitrogen (N), phosphate (P2O5) and potash (K2O) both as manufactured product and that supplied in organic manures. Further inputs of pesticides are made in order to prevent or limit crop losses associated with diseases, most notably potato blight, weeds and pests, including aphids that can transmit plant viruses. With these inputs together with land management practices, including cultivation and irrigation, mean that there is significant potential for pollution associated with growing potatoes.

DAP (Diammonium Phosphate) and NPK (Nitrogen, phosphorus and potassium) are among of the fertilizers used within the area with 46% P_2O_5, 18% N, 100% water soluble for DAP. They are excellent source of P and nitrogen (N) for plant nutrition, highly soluble and thus dissolves quickly in soil to release plant available phosphate and ammonium. A notable property of DAP and NPK is the alkaline pH that develops around the dissolving granule, temporarily increases the soil pH, but over a long term the treated ground becomes more acidic than before upon nitrification of the ammonium.

Study conducted by Defra 2009, Ferencz, L., Balog, A. (2010) and Stewart, W.M*et al.*, (2005)devoted that   Nitrogen and phosphorus increases the resistance of plants to adverse environmental factors (drought and frost), as well as diseases, but they are persistence to the environment and causes soil acidification, also Miguel Angel, (2004) document that, regular use of acidulated fertilizers generally contribute to the accumulation of soil acidity in soils which progressively increases aluminum availability and hence toxicity, but also common agricultural grade phosphate fertilizers usually contain impurities such as fluorides, cadmium and uranium in which they are potentially harmful impurities to the environment and impacts chronic health effects to human Example Severe damage of renal.

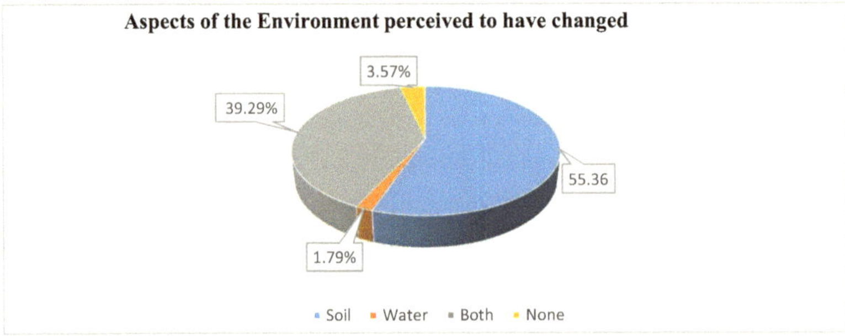

Source: Field data 2014

Figure 5: The Aspects of the Environment perceived to be impacted due to potatoes production at Kibohelo

4.8 Farming practices used in Kibohelo, trends, use of improved seeds, pestcides and the existence of value chain in potatoes production at the kibohelo village.

The production of potatoes in the area tend to increase as 58.93% of the farmers remarked, in which this is due to methods or practices used in potato cultivation such as crop rotation (28.57%), mulching (21.43%), miraba (16.07%) and small scale irrigation (19.64%) as it was observed in the area as well as it was mentioned by the farmers, but also the use of improved seeds as 48 (85.71%) of the farmers commented that they use them, while only 8 (14.29%) commented that they don't use improved seeds, in addition to 51 (91.07%) of the farmers commented that they use pestcides where only 5 (8.93%) they commented that they are not using pestcides. Where via this results indicates good conditions for the growth of the potatoes and existance of value chain of potatoes within the area as 56 (100%) of the farmers stated the presence of value chain of potatoes in the area, however 41.07% point out that production decreases, this is from the farmers who do not use improved seeds, others they lack enough time to look for their fields due to the reason that in many rural areas, agricultural activities alone cannot provide sufficient livelihood opportunities as documented also by Craig *et al.*, 2001. This is in terms of irrigating the crops, poor application of herbcides and pestcides as well as the chemicals to remove fog over the crops as the results in **Table 8** demonstraits.

These findings are in agreement with the observation made by Gildemacher*et al.*, (2009) and Gigliotti, C., Allievi, L. (2001) that the low yield of round potato is attributed to diseases, pests and poor quality of seeds used.

One of the insecticide used in the area is Dimethoate, which is a systemic and contact insecticide and acaride, effective against red spider mites and thrips on most agricultural and horticultural crops. Dimethoate might have carcinogenicity, birth defects, reproductive toxicity and mutagenic effects.

Table 8: Farming practices, trends, use of improved seeds, pestcides and the existence of value chain in potatoes production at the kibohelo village

	Kibohelo	
Farming practices used in potato production:-	**n**	**%**
Crop rotation	16	28.57
Small scale irrigation	11	19.64
Agroforestry	3	5.36
Mulching	12	21.43
Terracing	5	8.93
Miraba	9	16.07
Total	**56**	**100**
Increase/decrease in potato production :-		
Increase	33	58.93
Decrease	23	41.07
Total	**56**	**100**
The use of Improved seeds:-		
Yes	48	85.71
No	8	14.29
Total	**56**	**100**
The use of Pesticides :-		
Yes	51	91.07
No	5	8.93
Total	**56**	**100**
Existence of value chain of potato:-		
Yes	56	100
No	-	-
Total	**56**	**100**

Source: Field data 2014

CONCLUSION AND RECOMMENDATIONS

5.1 Conclusion

In this study, impacts of potato productions were found to be changes on the soil quality including effects on soil fauna and their processes as 55.36% of respondents remarked but also water quality changes due to contaminations from pesticides and synthetic fertilizers used during production of potatoes as 1.79% of the respondents commented. There are many methods or practices used in potato production but most of the farmers (28.57%), were seen to use either crop rotation or mulching (21.43%), because they seems to result in more yields, while the small number use agroforestry (5.36%) because it requires larger area and only few people own large cultivation lands in the area.

Despite awareness on environmental changes due to potatoes production by farmers still low level of awareness from the rest (21.43%) could escalate environmental degradation, sources of information are still limited and needs to be diversified. It was also observed that most of the farmers (91.07%) use synthetic fertilizers of which pose threat into the soil fauna components.

5.2 Recommendations

- Further detailed study of impacts of potatoes production especially on soil microbes in these fragile environments should be carried out in order to have a good understandings of our environment and sustainable cultivation activities.

-Sustainable agricultural innovation platforms should be introduced whereby all players along the value chain of the potato production and marketing must be effectively involved and committed.

-Training arrangements should be provided at village and district to ensure increasing level of awareness, utilization of and sustaining generated information.

-Effective utilize available research findings on afforestation, mulching, planting of different crop varieties as well as switching from farm to non-farm activities should be highly encouraged.

-Kibohelo local government should formulates and implements policies that would accelerate environmental conservation in the village through scaling up budgetary allocations to issues such as construction of improved irrigation schemes and capacity building to farmers to avoid practicing measures that lead to impact the environmental aspects in the area.

ACKNOWLEDGEMENTS

First and foremost I would like to thank my GOD, the almighty for making my studies and my academic years at Sebastian Kolowa Memorial University possible.

I would like to take this opportunity to thank numerous people who assisted me in getting this research done. Special thanks go to my supervisor Mr Fanakea Mwihomeke who tirelessly guided me throughout this research. I particularly thank him for providing me with all necessary reading materials.

I also wish to extend my gratitude to the Department of Science and all entire staff of Sebastian Kolowa Memorial University for their active roles they played in the development of this research. Last but not least, I would like to thank my family and friends for their support and prayers. Special thanks go to my father for his great assistance and financial support, and also to his great contribution in preparing this research.

It is not easy to mention all individuals who have contributed to the success development of this work. The few mentioned individuals will stand as a tower for all people who have been generous and supportive to see this study a success. To all of you I say thank you very much.

REFERENCES

Alcamo, J. 2001. Scenarios as tools for international environmental assessments. Environmental Issue Report no. 24. European Environment Agency, Copenhagen.

Andersson, J. A. 1996. Potato Cultivation in the Uporoto Mountains, Tanzania: an analysis of the social nature of agro-technological change. African Affairs 95 (378), 85–106.

Bank of Tanzania (2010). Monthly Economic Review, September 2010.

Blanken J, von Braun J, De Haen H (1994). The triple role of potatoes as a source of cash, food, and employment: effects on nutritional improvement in Rwanda. In: von Braun J and Kennedy (eds) Agricultural Commercialization, Economic Development, and Nutrition. The Johns Hopkins University Press, Baltimore and London, pp. 276-294.

CentroInternacionaldela Papa (CIP) (2008). The International Year of Potato http://www.cipotato.org/pressroom/facts_figures/2008_internationalyear_of_the_potato.a sp site visited on November 6, 2013.

Defra (2009) Protecting our water, soil and air. A code of good agricultural practice for farmers, growers and land managers. 124pp.

Duze M. and Mohamed ZI: 2006. Male Knowledge, Attitudes, and Family Planning Practices in Northern Nigeria. African Journal of Reproductive Health.

European Environment Agency 2006a. How much bioenergy can Europe produce without harming the environment? EEA Report No 7/2006. Copenhagen.

European Environment Agency 2006b. The integration of environment into EU agriculture policy – the IRENA indicator-based assessment report. Copenhagen.

Eze CC, Ibekwe UC, Onoh PO. And Nwajiuba CU: 2006. Determinants of adoption of improved cassava production technologies among farmers in Enugu State of Nigeria. Global Approaches to Extension Practice.

FAOSTAT, 2009. Production Crops, from http://faostat.fao.org Retrieved November 7, 2013

FAOSTAT, 2012. Production Crops, from http://faostat.fao.org Retrieved November 7, 2013

Ferencz, L., Balog, A. (2010): A pesticide survey in soil, water and foodstuffs from central Romania. Carpathian Journal of Earth and Environmental Sciences, April 2010, 5(1), pp 111-118

Gigliotti, C., Allievi, L. (2001): Differential effects of the herbicides Bensulphuron and Cinosulphuron on soil microorganisms. Journal of Environmental Science and Health, Part B: Pesticides, Food Contaminants, and Agricultural Wastes 36(6), pp 775 - 782.

Gildemacher PR, Demo P, Barker I, Kaguongo W, Woldegiorgis G, Wagoire WW, Wakahiu M, Leeuwis C, and Struik PC (2009). A Description of Seed Potato Systems in Kenya, Uganda and Ethiopia. American Journal of Potato Research (2009). 86 (5): Publisher: Springer New York: 373-382

Craig, C. and Gordon, A (2001), Rural non-farm activities and Poverty alleviation in Sub Saharan Africa. Policy Series 14, Natural Resource Institute University of Greenwhich.

Junge B, Deji O, Abaidoo R, Chikoye D. and Stahr K: 2009. Farmers' Adoption of Soil Conservation Technologies: A Case Study from Osun State, Nigeria. The Journal of Agricultural education and extension.

Lupilya GS: 2007. Assessment of social support projects for vulnerable groups towards poverty reduction: A case study of TASAF in Bukoba district. M.A Dissertation, Sokoine University of Agriculture, Morogoro, Tanzania. 112 pp.

Miguel Angel, (2004). Biodiversity and pest management in Agroecosystem, 2nd Edition, Psychology press.

Minoe, S., Barker, D., Dixon, J (2003), Status of Farm Data system and Farmer Decision Support in Sub- Saharan Africa, FAO Rome.

Okoedo-Okojie DU. and Onemolease EA: 2009. Factors affecting the adoption of yam storage technologies in the Northern Ecological zone of Edo State, Nigeria. Journal of Human Ecology.

Prakash, A. 2010: The role of potato in developing country food systems. In Cromme N., Prakash, A., Lutaladio, N. &Ezeta, F (Eds.), 14–24.

Stewart, W.M.; Dibb, D.W.; Johnston, A.E.; Smyth, T.J. (2005). "The Contribution of Commercial Fertilizer Nutrients to Food Production". *Agronomy Journal* 97: 1–6.

Toulmin, C. International Institute for Environment and Development. Food security in 2050: how can we make it fairer and more sustainable http://www.iied.org Retrieved

December 16, 2013 from how can-we-make-it-fairer-and-more-sustainable

Uyole Agricultural Research Centre (UARC) (1990). Kilimo Bora cha ViaziMviringo. Extension Leaflet No. 49. Shirika la KilimoUyole, Mbeya, Tanzania.

Wichelns, D., & Oster, J. D. (2008). Sustainable irrigation is necessary and achievable, but direct costs and environmental impacts can be substantial. Agricultural Water Management, 86, 114-127.

Williamson, S., Ball, A., & Pretty, J. (2008). Trends in pesticide use and drivers for safer pest management in four African countries. Crop Protection, 27, 1327-1334.